Operator's Guide

TO CENTRIFUGAL PUMPS

Operator's Guide

TO CENTRIFUGAL PUMPS

What Every Reliability-Minded
Operator Needs to Know

VOLUME 2

ROBERT X. PEREZ

Library of Congress Control Number:		2014921849
ISBN:	Hardcover	978-1-5035-2501-6
	Softcover	978-1-5035-2502-3
	eBook	978-1-5035-2503-0

Print information available on the last page.

Rev. date: 09/25/2015

To order additional copies of this book, contact:
Xlibris
1-888-795-4274
www.Xlibris.com
Orders@Xlibris.com
638081

TABLE OF CONTENTS

DEDICATION

I would like to dedicate this book to my wife, Elaine. Without her support, this book would not have been possible.

ACKNOWLEDGEMENTS:

I WOULD LIKE TO thank:

1. Elaine Perez for proofing this book and offering valued advice on the student exams.
2. Dan Ellwood of Chesterton Seals for providing the mechanical seal images used in the exams.
3. Sonny LeBleu for helping write Chapter 1 of this book.

More Ideas on Centrifugal Pump Reliability

MANY READERS HAVE found Operator's Guide to Centrifugal Pumps (Xlibris, 2008) to be a valuable training resource for process operators. Volume 2 provides additional training material in the form of student challenge exams and additional exposure to reliability topics.

Volume 2 has two goals: The first goal is to continue to build on the general theme of pump reliability in process facilitates. The second goal is to provide student challenge exams for those wanting to master the material in Operator's Guide to Centrifugal Pumps.

The first chapter covers how process operators can be better utilized to improve pump reliability. The importance of operating training and commissioning will be covered in the next two chapters. In the final chapter, a methodology for addressing troublesome centrifugal pumps is presented.

In the back of this book, the reader will find three different student challenge exams for those wanting to master the material in *Operator's Guide to Centrifugal Pumps* (Xlibris 2009). Answers to all the exam questions are provided, along with their location, within *Operator's Guide to Centrifugal Pumps* (Xlibris 2009). There are a total of 150 challenge questions with their answers.

I hope the reader finds this additional content useful.

Robert X. Perez

CHAPTER 1

Operators: An Underutilized Resource

Figure 1.1, An operator keeps an eye on key process parameters
that are displayed in a control room

OPERATORS ARE THE eyes, ears, and manipulators of
switches and valves in their plant. Typically, they receive
training either from a 2 year associates degree or plant on-the-job
training. Their training may or may not include the understanding of

how their equipment works, or the significance of pump performance curves. I venture to guess that to most operators, the mechanical seal is simply a "black box" that is not well understood. Operators may not need to know precisely how a mechanical seal works, but they need to know the seal is the major reason pumps are taken out of service.

Most operators are expected not only to identify there is a problem with a piece of rotating equipment but to troubleshoot the problem as well. They have been taught what to do to start, operate, and shutdown the plant, but without the understanding of how the equipment is supposed to work and the basics of a pump curve, troubleshooting will be very difficult to perform. Most operators do not feel they have control over the process, but that they are subject to its whims. For example, operators know how to detect when a pump is cavitating; but they do not feel they have any control over the cavitation and so they might identify it but feel no obligation to do anything about it. They do not realize they do have some limited control over the phenomenon. For example, if the pump takes suction from a tank, they need to ask: Can the tank level be raised? Can the tank be pressurized? Can the flow be decreased to lower the NPSHR (net positive suction head required)? These are some of the parameters that can be controlled by operators and may possibly eliminate the cavitation phenomenon completely. Remember that if you are going to transfer more liquid with a pump that you must increase its suction pressure in the form of a higher NPSHA (net positive suction head available) condition.

Teach look, listen, feel

Most companies believe that operators are already using their senses effectively and perhaps they are. It is also important for the operator to consider what is causing the evidence that they are sensing. For example if there is a black mark on a tank near a hose, don't just notice the black mark but ask why is that mark there. If this is not done, it is possible that a hose is rubbing on the tank and may eventually lead to a hole in the hose, preventing the use of the equipment the hose is connected to. Many problems can be detected before failure occurs by using simple monitoring techniques, but the operator should keep in mind that when these clues are detected, it means that failure is very near. The look, listen, and feel (LLF) method is helpful for gathering data and then determining and implementing a permanent solution to the problem before a costly failure occurs. The LLF method is also helpful in allowing the operators to learn about their equipment and what it is like when they are operating well or when there is a problem.

Pump curves

Operators need to be taught the basics of pump performance curves. They need to know the relationships between pressure, flow, horsepower and net positive suction head (NPSH). These relationships need to be taught for both centrifugal and positive displacement pumps. The ability to troubleshoot is extremely limited if pump performance relationships are not well understood. Many operators know that the discharge pressure on a centrifugal pump will rise if the discharge valve is closing. They also, wrongly, think that the pump is drawing more horsepower under these conditions, which is because they don't understand the relationships between horsepower and flow mentioned earlier.

Pump curves can help explain why the act of pumping more product than the pump is rated for means more net positive suction head will be required to prevent cavitation.

Most operators do not understand these relationships and so therefore don't feel that they have any control over the potential for cavitation in their pumps. For example, most operators don't know that they can reduce the likelihood of cavitation by raising a pump's suction pressure by raising tank level or raising the pressure in the suction vessel. They normally don't think about reducing the output from the pump to reduce the net positive suction head required. Operators need to be taught that if a pump has a bypass to allow the pump to operate at various flows and head, they can close or reduce the spillback flow to reduce the NPSHR of the pump and possibly reduce or eliminate cavitation. Another concept that needs to be explained to operators is that cooling the incoming liquid is another way to reduce the possibility of cavitation, but generally this is usually a more difficult correction to implement.

Teach Root Cause Failure Analysis (RCFA) Methods

The simple "why" method of root cause analysis is a great help in identifying the root cause of a problem and is something that can be employed by the operator at the time a problem is discovered. Many operators think their job is to identify a problem and then turn it over to someone else to solve. This may be true, but since the operator is the person looking at the problem, he or she can gather data on the spot so that a permanent solution may be readily identified.

Operators and capital projects

To improve the reliability of new installation, operators as well as mechanics should be included in the capital projects process, at least at the beginning of the design and equipment selection process. Operators and mechanics can be helpful in assessing the operability of the proposed installation, and the mechanics can help determine the space required for proper maintenance to take place. For example, many engineers do not have enough experience to realize the space requirements to operate a 24" valve. Mechanics understand that there must be enough room to pull heat exchanger bundles, while allowing room to perform maintenance on rotating equipment.

No one knows what the operator knows

Figure 1.2, Operators know what their rotating equipment sound like and feel like when they are operating normally

Operators are the ones who know what the equipment sounds like and feels like when it is operating as it should. The best mechanics or engineers do not know what the operators know. The operator is the one that can answer the question "what was this piece of equipment doing yesterday, last week, last month, etc." No one else is as close to the machine as the operator is. The operator is also the one who can associate an equipment problem with a particular plant activity. For example, an operator might realize that a pump bearing fails two weeks after a plant wash down, or that after plant process upset a certain piece of equipment has a reduced output.

The operators are capable of giving process specific information, such as the process requires that both main and spare pumps must be on line to make normal production rates. This operating requirement means that this new normal situation has eliminated your spare, i.e. you no longer have a spare pump in the event of a pump failure. Furthermore, if the two identical pumps do not perform identically, the operator is the individual that can identify these deficiencies and alert engineering to the problem.

It is a good idea to allow operators to make a first pass at troubleshooting their equipment. Once they are given the tools and knowledge required, they will become an asset in the resolution of chronic field problems. Most operators do not have the training of how the equipment works, so it is difficult for them to know when it is not performing as designed. They only know that there is a problem because the equipment is not performing as it once did or as would like it to perform. The second part of using these vital individuals as field troubleshooters is to allow them to make mistakes. If operators are chastised after one mistake, they will get "gun shy" and not attempt to troubleshoot or make decisions based on their observations. They must be allowed to make mistakes as long as they do not continue to make the same mistakes over and over. The mistakes are how we all learn and improve our observation and troubleshooting skills.

Author's note: For more information on the "look, listen, feel" philosophy for operators you can refer to "Operator's Guide to Rotating Equipment" (AuthorHouse Publishing 2014). The goal of this new book is to provide useful ideas and advice aimed at empowering plant operators whose job it is monitor critical rotating equipment. Julien LeBleu is the primary creator of the "look, listen, feel" philosophy of monitoring machinery.

ROBERT X. PEREZ

CHAPTER 2

Operator Training is a Key Factor in Centrifugal Pump Reliability

Figure 2.1, Operator training is an inexpensive means of improving and maintaining pump reliability at your site

New Pumps = Operator Training

THE WAY A pump is started up and operated has a profound effect on its overall reliability. During its lifetime, a pump must be started and shutdown numerous times and operated at various degrees of hydraulic load. To ensure every pump attains its expected serviceable lifetime, it must be properly vented and provided with adequate suction pressure, its seal or seals must be lined up properly at all times, and it must be started up with sufficient back pressure to prevent pump runout. All these operating requirements must be learned and then practiced repeatedly until they become second nature.

After new pumps are purchased and installed, the commissioning process begins. (Commissioning is the process of checking an installation to ensure everything is installed and that the pump, driver, and auxiliaries are ready to start.) It is a mistake to believe all pumps are alike and that every operator knows how to operate every type of pump. A safer viewpoint is to assume that every pump is unique and requires special knowledge to be properly operated and monitored.

An important step in the commissioning process is training the operators about any new pumps that have been installed. Operators need to know:

1. What types of pumps are being installed?
2. What type of seals the pumps have?
3. What type of lubrication system is being used?
4. What are the basic operating properties of the pump?
5. Whether there are any unique operating requirements, such as venting, preheating, minimum flow requirements?
6. What they need to look for as indications that something is wrong.

This specialized information cannot be acquired through osmosis. It must be transmitted through some type of training.

Learning it Wrong

One day I was talking to a seasoned operator about how he prepared pumps for service after repairs. I asked him specifically how he vented some of the critical process pumps that handled environmentally sensitive fluids. He told me he didn't have to vent them at all because they were all self-venting. I stood there for a few moments shocked, bewildered, and then disappointed. I soon understood what it was he was telling me. Whoever had trained him long ago taught him he never had to vent pumps before startup because they only purchased self-venting pumps. He was right that most pumps are designed to be self-venting but the associated pump piping is not!

If a pump's piping is fully vented, most pump casings are designed so that they will air-free themselves. But, if the pump's piping is not properly vented, the pump can have difficulty eliminating trapped air in the casing and piping system subsequent to a repair. In extreme cases, a pump can destroy itself before becoming fully primed if not properly air-freed. This sad story illustrates the problem with on the job training: Misunderstandings, misinformation, and fallacies are often perpetuated from one co-worker to the next until they are cleared up with some type of training. There is also the real possibility that what was once learned has either been forgotten or is now out of date.

On the Job Training versus Formal Training

I don't mean to be overly critical of this operator or the company he works for, but this example begs the question: Is on the job training (OJT) enough for pump operators? Processing plants are complex systems that utilize a wide range of equipment and processes. Plant managers must judge how their precious training dollars are best spent. They must also deal with a continuous stream of transfers, retirements, and resignations.

Unfortunately, there are probably countless operators that have never been provided with any formal instruction. This is probably because pumps are often considered less important due to the fact that they are often spared and relatively inexpensive. When they fail, people shrug their shoulders and say, "It's just a pump." There are operators who have been taught simply to push the start button and walk to their next task. It's only when a major pump-related event occurs that management gets interested in training. The best way to avoid major or frequent failures in the first place is to initiate and maintain a formal pump training program.

Here are several points to consider about the lowly pump:

1. Pumps are typically installed without any sort of automation, i.e. flow control valves, automatic venting valves, automatically opening suction and discharge valves, etc. If no automation is installed, it is up to the operator to know how to monitor and protect the pump from unsafe conditions.
2. Often pressure gauges or flowmeters are absent to allow for monitoring. The hope is that, when the start button is pushed, product will flow at the desired rate in the proper direction. How can a pump be monitored without this instrumentation?
3. Pumps are very sensitive to start-up conditions. They don't like to be operated without liquid or without back pressure.
4. One size does not fit all when it comes to pump start-ups. The wide variety of pump types in a plant necessitates that operators understand what start-up procedure is required for a given pump design and situation.

Training Economics

How much training is required? Let's first ask the following question: How much are your pumps costing you? To answer this query, I recommend you figure out how much your pumps are costing you per operator. Let's say you have 1000 pumps, with a 36 months MTBR, costing $6000 per repair. If your facility employs 100 operators, you are spending $20,000 per operator (see calculation below). The bet is that training will improve your MTBR and reduce the cost per repair due to the elimination of secondary damage from improved monitoring. Improving the MTBR by 2-5% alone is worth $400 to $1000 per year per operator. Elimination of secondary damage, releases, and fires represent addition benefits.

Calculation of the Cost of pump repairs per operator

$$\frac{Cost\_of\_repairs}{Operator} = \frac{N \times t \times \$_R}{MTBR \times N} = \frac{1000 \times 12 \times 6000}{36 \times 100} = \$20000$$

In this case, it seems well justified to budget $400 to $1000 per year per operator. This provides a one year payback on your investment. Remember that your 2 to 5% improvement will continue well past the training year. Note: When your MTBR exceeds 8 years you can begin backing off some your training budget.

I don't mean to imply that pump training will solve all your problems. There are many reasons why pumps fail and most of them are unpreventable by operator intervention. Here are a few:

- Normal wearout, i.e. end of life
- Hydraulic misapplications
- Improper repairs
- Wrong materials of construction

However, I feel it is safe to say that correct operating procedures, proper startups, and careful monitoring will undoubtedly <u>extend</u> pump operating intervals and <u>reduce</u> repair cost by avoiding secondary damage with them. Additionally effective training will not only significantly reduce operator related failure it will also instill more confidence into your operators.

Training Options

Figure 2.2, Knowledge is not luxury; it's a requirement for operating success

There are a number of proven ways to educate operators. Here are a few:

- **Formal training in a classroom setting.** You can bring in a consulting firm or pump supplier to provide your operators with classroom training composed of theory, proven reliability methods, and hands-on demonstrations. Self-paced training, either online or with study guides, is another proven option. There is a full array of self-directed training to be found on the web. Formal training should probably be provided <u>every five years</u> to ensure exposure to new technology and concepts.
- **Refreshers provided in-house.** You can also enlist the assistance of in-house machinery engineers or technicians for basic refresher training. This type of training should be provided <u>every other year</u> to ensure on-going technical competence.
- **Hands-on "practicals".** Senior operators can watch and evaluate junior operators as they start-up pumps. Yearly "practicals" and

field demonstrations should be encouraged. A great example of field training would be to have a junior operator swap from a main to the spare pump while a senior operator is watching.

- **Frequent reading of technical journals or textbooks**. This type of training should be provided to your operators on a continuous basis. Ensure there are enough copies of current pump journals and textbooks in your shops and control rooms for all to peruse.

Remember that operators are the hands, eyes, ears, and noses of our processes. They are vital to profitable and reliable operations. It's our choice; we can either set them up for certain failure or prepare them for operating success. Training is one of the most inexpensive means I know of improving overall process reliability.

CHAPTER 3

Crucial Moments in a Centrifugal Pump's Lifetime

Figure 3.1, Double checking pump design and installation details will pay off in a lifetime of reliable operation

The Devil's in the Details

THERE ARE MANY detail steps involved in the selection, installation, and commissioning of a centrifugal pump. These steps involve the equipment selection process, the piping system design, and the installation. No detail is too small to consider unimportant. Collectively, this overall process will either ensure a successful installation or a lifetime of maintenance and operational headaches. Many individuals must perform their jobs correctly to ensure the overall success of a pump installation.

I recently saw a commercial on TV touting an air conditioning system, where the announcer stated, "The most important day of an air conditioning system's life is the day it's installed." I thought about this a lot and decided that this is not true for pumps, where there are many important days in a pump's life. If you want a perfect pump installation here are days I think are truly critical:

- The day pump hydraulic requirements are specified by the process engineer
- The day the pump specifications are written
- The day the pump is selected
- The day the mechanical seal is selected
- The day the piping and control system are designed
- The day the pump is installed

Additionally, if you want to reap the benefits of your "near perfect" pump installation, you must also be aware of these important days:

- Every day it's repaired
- Every day it's started up

In reality, there can never be a perfect pump installation. Process engineers never know exactly what flow range or pressure rise will be required during the life of the pump. Processes tend to evolve due to market forces and technological improvements, forcing pumps to operate away from their sweet spots. Also, pump manufactures don't always have the exact pump to fit your needs. To make things worse, tight project economics can severely limit the level of design features your installation is allowed to have.

Thinking about all these selection forces and considerations can give a pump user a severe case of anxiety. How can we ever hope to obtain a perfect pump installation? There is no definitive answer to this impossible question. All I can provide are a few guidelines based on many years of experience in the pump trenches:

1. Have the process engineer specify the possible range of flow and head requirements, not just maximums. Minimum flow rates for start-ups or other special conditions may require spillback lines or variable speed drives for pump protection.
2. Have the process engineer specify the possible range of process temperatures and if the stream is expected to be dirty or corrosive. This input will allow for a suitable mechanical seal selection and flushing plan.
3. Follow the PIP guidelines when designing the pump's piping system.
4. Select a control system that allows your pump to operate at the lowest stress level, i.e. near the best efficiency point, most of the time. Beware of temperature and level controls that can force your pump to operate well away from the pump's BEP.
5. Also diligently follow PIP installation guidelines to ensure your pump is installed relatively stress free and rigidly grouted in.

After the pump has been purchased and installed, remember that the road to pump reliability must run through your training department. Everyone involved in operating the new pumps need to be trained to think about pump reliability and learn how to put their knowledge into practice. They need to understand that every start-up, shut-down, and inspection is an opportunity to improve their process' reliability.

This goes for your mechanics as well. Instill in them that they are not just parts changers, they are also reliability technicians. Every pump failure and subsequent repair is an opportunity for improvement. Frequent discussions between the operators, mechanics, and machinery engineers should be encouraged as a means of solving problems related to detrimental operating and repair practices.

I'll leave you will this quote attributed to Admiral Hyman G. Rickoever, U.S. Navy, known as the "Father of the Nuclear Navy".

"The devil's in the details—but so is salvation"

Details matter, especially during the most crucial days of a pump's life. Remember that pump reliability is everybody's job, not just the machinery engineer's.

CHAPTER 4

When Good Pumps Turn Bad

A Methodology for Addressing Troublesome Pumps

Figure 4.1, Photos of some Troublesome Centrifugal Pumps

Bad Actors

WE ALL HAVE them. They cause us to worry incessantly, lose sleep, and frequently miss precious time with our families. They are often the bane of processes that require liquids to be transferred from one location to another. These mechanical monsters are pumps widely called unflatteringly "Bad Actors." By definition Bad Actors are pumps that fail so frequently that they stand apart from the rest of the pump population. I've known of bad actors that have failed as many as 18 times in one year. These troublesome machines sap precious resources from our maintenance departments, preventing us from achieving world-class reliability performance.

I am sure that these troublemakers were all carefully selected by well-intentioned vendors and project engineers and installed dutifully by construction companies. But the devil is in the details. Fatal flaws, ranging from flimsy shafts to poor operating practices, crept into these pumping systems, undermining their reliability and compelling them to lead notorious lives.

The inordinate number of failures experienced by Bad Actors tends to dramatically skew the mean time between repairs (MTBR) for a plant downward. For this reason, a key strategy for improving your plant's MTBR is to start by identifying and improving the reliability of your most infamous pumps. Here, I will present a straightforward methodology for addressing a site's most problematic pumps.

Methodology for Addressing Bad Actors

1. Define, list, and compare

To address Bad Actors, you must first define what they are. Usually definitions contain a combination of failure rate and repair cost criteria. For example, you may define any pump that fails two (2) or more times and results in more than $10,000 in repairs costs over the previous twelve (12) month period as a Bad Actor. (Note: These criteria can be modified to satisfy management preferences.)

Some plants also include lost production losses during the same reporting period. You can simplify your reporting by combining repair cost and production losses into a single figure called "losses." These multiple criteria tend to cull nuisance pumps that fail numerous times per year but do not have a large annual repair total. By using the multiple criteria of failure rate and repair costs, you can quickly identify the pumps having the greatest impact on reliability.

2. Go after the top Bad Actors

After creating a list similar to Table 4.1 below, you simply sort in descending order of the most costly to least costly pump. The top ten on this list represents your Bad Actors. This list should probably be compiled quarterly, semiannually, or annually. Once you have your hit list, you start by attacking the worst of the worst actors.

Table 4.1, List of Bad Actors

Pump #	# of Failures in the last 12 months	Repair cost over the last 12 months	Production Losses over the last 12 months	Total Losses in the Last 12 Months
12	2	$5,000	$250,000	$255,000
4	3	$25,000	$120,000	$145,000
18	5	$8,000	$50,000	$58,000
16	3	$50,000	$0	$50,000
19	3	$40,000	$0	$40,000
20	2	$15,000	$14,000	$29,000
14	6	$24,000	$0	$24,000
11	3	$20,000	$0	$20,000
10	5	$13,000	$0	$13,000
3	3	$12,000	$0	$12,000
8	4	$12,000	$0	$12,000
1	3	$5,000	$5,000	$10,000
9	2	$7,500	$0	$7,500
13	2	$7,500	$0	$7,500
7	6	$6,000	$0	$6,000

ROBERT X. PEREZ

6	2	$5,200	$0	$5,200
2	2	$3,000	$0	$3,000
15	1	$2,500	$0	$2,500
5	4	$2,000	$0	$2,000
17	4	$1,500	$0	$1,500

3. Examine the equipment history

The next few steps describe how to evaluate each Bad Actor on your hit list. Let's examine a hypothetical data set for a Bad Actor. To construct a data table similar to Table 4.2, you need to know the 1) date of each failure and 2) the repair cost for every past failure in the time frame of interest. You also need to have a starting point defining where your clock starts.

In the example below, the first failure occurred 15 months after your defined starting time and the cost of the repair was $5000. The next failure occurred 12 months after the first failure and resulted in a repair cost of $5500. This means that the cumulative time (*3rd column from the left*) for the second failure was 27 months and the cumulative repair cost (*5th column from the left*) for the second was $10,500. So for each subsequent failure you keep accumulating the failure numbers, time, and repairs costs, as seen in the cumulative failure, time, and costs columns below.

Table 4.2, Hypothetical Bad Actor Data

Cumulative Failures	Time Since Last Failure (months)	Cumulative Time (months)	Repair Cost	Cumulative Costs
1	15	15	$ 5,000	$ 5,000
2	12	27	$ 5,500	$ 10,500
3	18	45	$ 6,000	$ 16,500
4	16	61	$ 4,500	$ 21,000
5	19	80	$ 5,000	$ 26,000
6	12	92	$ 3,500	$ 29,500

7	20	112	$ 6,500	$ 36,000
8	16	128	$ 5,000	$ 41,000
9	19	147	$ 5,200	$ 46,200
10	12	159	$ 4,000	$ 50,200
11	9	168	$ 3,500	$ 53,700
12	6	174	$ 6,000	$ 59,700
13	5	179	$ 6,500	$ 66,200
14	4	183	$ 8,000	$ 74,200
15	6	189	$ 8,500	$ 82,700
16	5	194	$ 7,500	$ 90,200
17	4	198	$ 9,000	$ 99,200
18	8	206	$ 10,000	$ 109,200
19	7	213	$ 9,000	$ 118,200
20	6	219	$ 8,500	$ 126,700

If you plot the cumulative failure number and cumulative repair cost value versus the cumulative time, you will get a plot like the one shown in Figure 4.2. I call these <u>reliability growth plots</u> because they clearly illustrate if the failure rate is constant or changing over time and if the rate of cost to perform maintenance is changing over time. A constant slope means the failure rate is constant, while a curving plot means the failure rate is changing. The reliability growth plot in Table 4.1 shows a constant failure rate up until month 160 to 170. After that time, the failure rate and expenditure rate begin to increase and eventually settle into a new higher failure rate.

Figure 4.2

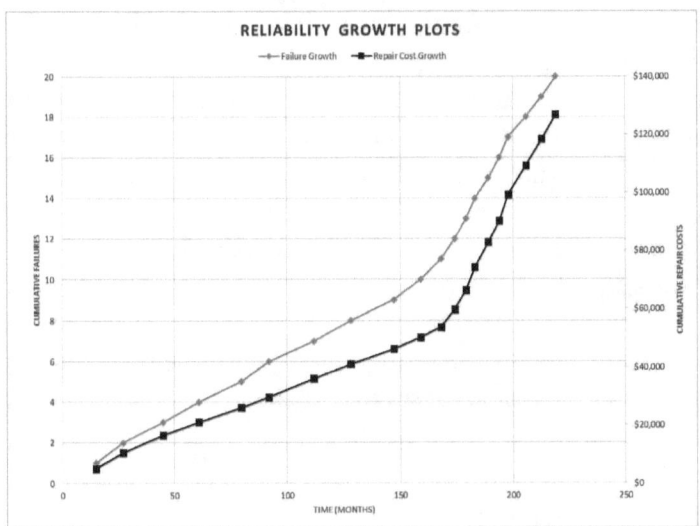

Figure 4.2, Reliability Growth Plot for Hypothetical Pump

These reliability growth plots offer a wealth of information. First, the cumulative failure plot shows you if the failure rate is constant or changing with time. If the failure rate did change, it tells you when the change occurred. You can discover if the failure rate was always bad or if it changed at some time in the past. Similarly, by examining cumulative repair cost data, you can determine if something changed in the past or if failure costs have been constant from the beginning.

If there is a defining moment when reliability went south, you can then ask: "what changed?" You can interview operators, mechanics, etc. to find out if there is a reason for the observed change in reliability. Field personnel represent a treasure trove of information. Many a time has an operator or mechanic provided key insights that assisted me in complex RCFA's. You may discover the:

- Nature of the process has changed
- Control scheme was modified in the past
- Seal flush source was modified due to process contamination concerns

Interviewing those close to the equipment under investigation is a great way to uncover subtle issues that may be affecting reliability performance.

Here is a telling example involving pumps that were failing every few months. We discovered that a production engineer decided to eliminate the use of an external seal flush because he felt it was contaminating the process. After we convinced him to reinstitute the flush at a lower, friendlier rate, seal life returned to an acceptable level.

If the general trends of your reliability growth plots are fairly constant over the operational lives of the pumps in question, then you can assume there is something wrong with the basic design of the pumping system. Possible causes may include:

- Poor installation
- Undersized pump shaft
- The wrong pump selection
- The wrong mechanical seal selection
- The wrong seal flush arrangement
- Excessive piping strain

The reliability growth plots also tell you how much the pumps are costing. In our example, we can quickly conclude you have spent $126,700 over a period of 219 months. This equates to an annual rate of $6942. The annual rate of expenditure is very important because this tells you the value of solving the problem. In this example, if I assume I can reduce my repair cost to 25% of the starting value, I can only hope to save about $5200 per year. So if I require a 2 year payback to justify capital expenditures, then I can only spend about $10,000 on a solution.

To ensure I have an acceptable return on investment, I try to avoid working on pumps that have annual repair and process losses less than $10,000. I have found that economic justification of reliability projects for pumps with annual losses less than $10,000 is next to impossible, unless they involve simple seal improvements, bearing upgrades, or procedural changes.

4. Conduct detailed design audits

Figure 4.3, Performing field audits can help identify
subtle issues affecting pump reliability

The next step requires that a design, installation, and performance audit be conducted, which involves:

- Reviewing the 1) pump selection, 2) driver selection, 3) seal design, 4) piping design, 5) control system design, etc.
- Conducting a detailed vibration analysis of the pump, motor, and piping system
- Reviewing of the baseplate and foundation design
- Assessing current hydraulic performance versus what is expected, etc.

This is your chance to ask: Is this is the correct pump and system design for your specific service?

I have found that a valuable tool that augments the design audit is the "cold eye review," which is an assessment of a system or process by an experienced, unbiased third party. The way this works is that you ask another pump engineer, technician, or operator to accompany you to the field and inspect the pump in question. Your instructions are for him or her to look for anything that might be considered unacceptable, such as 1) too much vibration, 2) a lack of piping supports, 3) absence of pressure gauges, etc., or 4) a lack of vents or drains.

Honest feedback from a third party can often bring valuable insights to the table. After living with a problem pump for a long period of time, we can become oblivious to issues right in front of you. The cold eye review can help uncover potentially important issues that you may have overlooked by all those actually living and working close chronic Bad Actors.

5. **Perform an RCFA**

Now that you have reviewed the failure history and conducted a design audit, you should begin to understand what's going on with your pump. The next analysis step requires that you conduct a root cause failure analysis to drive you to the root cause of the failures. Remember that you should not stop at a physical root cause, such as the pump failed due to a bearing failure or shaft failure. You must drive your investigation to uncover any the latent root causes lurking beneath the surface.

Whatever RCFA methods that is required or recommended by your reliability department will suffice. The key point here is that you must be open-minded during the investigation process. It is important that, if you employ a team approach in your RCFA, only select team members who are willing and able to keep an open mind through the process and avoid hidden agendas. Your main goal is to seek the true cause of your repetitive failures.

6. **Determine a path forward**

Once the root cause and contributing factors are established, it will then be time to formulate a recommended plan of attack. My philosophy is that less is more. In other words, it is easier to sell two recommendations to management than 20 recommendations. Furthermore, it is also easier to implement two recommendations than 20. This doesn't mean you cannot have more than two recommendations; I am simply suggesting that by only presenting the highest priority recommendations to management your chances of securing approval dramatically improves.

Don't be afraid to fail. We all fail occasionally. The only way I know that you can improve your batting average is to gather lots of data, analyze it in exhaustive detail, and perform detailed RCFA's with knowledgeable people. Remember that the RCFA process is a process of continuous improvement. Some problems are so complex that they may take several tries to solve.

7. **Track your progress**

Figures 4.4, Attacking bad actors will eventually have a positive effect of your sites overall pump reliability

After obtaining management approval, it is time to implement your recommendations in a timely fashion. This is where the rubber meets the road.

Your last step requires that you track benefits of your improvements and determine if your efforts bear fruit. The proof of your success will be seen in an updated reliability growth plot, where hopefully reliability improvement will manifest itself as flattening reliability growth plots.

If you do see a clear improvement, make sure to publish the good news. Management, operating personnel, and all those involved in the RCFA investigation need to know about the results of their efforts. Early successes will motivate everyone to continue whittling down the number of your bad actors until you reach your plant-wide MTBR target.

Summary:

Should we simply say that *unreliability happens*? Yes, there are seemingly insignificant decisions and events that can occur during the early life of a pump that eventually lead to subpar reliability performance. But we can also say that reliability improvements, if implemented properly, can turn things around. Successful reliability improvement programs require that latent root causes be identified and corrected. Starting with your most troublesome pumps, you must systematically whittle down your list until world class reliability is achieved.

Remember these critical steps in Bad Actor reviews:

- Define, list, and compare
- Go after the top Bad Actors
- Examine the equipment history
- Conduct a detailed audit
- Perform an RCFA
- Determine a path forward
- Track your progress

Remember that reliability is a journey not a destination. There will always be another pump failure to analyze and learn from. Every failure should be considered an opportunity to learn more about your equipment, processes, and systems and improve them.

CHAPTER 5

Centrifugal Pump Activities for Operators

1. What do you think the main duties of a process operator should be with regards to centrifugal pumps?
2. What would you like to learn more about regarding centrifugal pumps? Talk your answer over with your supervisor.
3. List at least three things you look for during your centrifugal pump inspection rounds?
4. Compile a list of pumps that have failed 2 or more times in the previous 12 months. Find out if there are any upgrades or modifications planned to improve their reliability.
5. What is the average lifetime of pumps in your area? Is the current MTBR acceptable to management? If not, try to form a MTBR improvement team and begin by attacking the most troublesome pumps first.
6. Select a critical process pump in your area. Find the pump curve and study it. Find the pump's: a) best efficiency point (BEP), b) normal flow rate, and c) rated horsepower.
7. Pull the equipment file of critical pumps in your plant. Have a machinery engineer explain how the seals, sealing system, and lubrication system work. Pass this information on to another operator.
8. Go to the field and find pumps with pressure gauges on them. See if any of the pressure gauges show unsteady pressures. If you find a pressure gauge indicating an erratic pressure, try to find a reason for the unsteady pressure.

9. Talk to a vibration tech about a pump vibration problem. Ask a lot of questions. Find out what the vibration tech thinks is the cause of the high vibration. Explain what you learned to another operator.
10. Compile a list of seal flush plans for a process unit. If you don't understand any of the seal flush plans in your area, discuss the ones you are unsure about with a mechanic or machinery engineer.
11. List all the pump lubricants in your operating areas. What is different about them? If you don't know, ask a machinery engineer to explain the differences between the lubricants.
12. Find an oil analysis report for a critical pump. Have a machinery professional explain it to you.
13. Witness an in-house disassembly and inspection of a pump in need of repair. Ask a lot of questions. Ask if the failure was a premature failure or if the pump had a normal run length. If the pump failed prematurely, find out why it failed and what can be done to prevent a repeat failure.
14. A centrifugal pump had been running normally for several months. Overnight, the pump's discharge pressure drops from 150 psi to 50 psi. What could be the cause of this loss of pressure? Explain your answer.
15. Immediately after the repair of a centrifugal pump, it is unable to generate any pressure. What could be the cause? Explain your answer.
16. Participate on a pump failure analysis team. Ask a lot of questions. Try to determine the root cause of the failure. Follow-up to make sure all the recommendations made by the team are implemented.
17. Participate on a mechanical seal failure analysis team. Ask a lot of questions. Try to determine the root cause of the failure. Follow-up to make sure all the recommendations made by the team are implemented.
18. Watch someone start-up a pump after a repair. Record your observations.
19. Watch someone switch from the main to the spare pump (or vice versa). Record your observations.
20. Teach someone how to start-up a centrifugal pump.

21. Visit an outside pump shop during a centrifugal pump repair. Ask a lot of questions. Find out why the pump failed. What did you learn?
22. Visit an outside motor shop during a motor repair. Ask a lot of questions. Find out why the motor failure. What did you learn?

Student Challenge Tests

Introduction

THE FOLLOWING CHALLENGE tests and answers are based on the material found in the first book of this series: *Operator's Guide to Centrifugal Pumps* (Xlibris 2009). These tests are intended to be used by students wanting to test their progress or by instructors wanting to assess student proficiency after they have read and studied the material in *Operator's Guide to Centrifugal Pumps* (Xlibris 2009).

It is my hope that by challenging students with these assessments they will gain a more thorough knowledge of centrifugal pump operations.

There are three tests, i.e. Test A, Test B, and Test C, included here to allow for several testing opportunities. You or the instructor can decide if the assessments are to be "open book" or "closed book." The goal is to assist the student in learning the book material by providing practical questions that reflect its key points.

All the answers can be found in the back of this booklet—along with detailed calculations, if applicable, and notes as to the location of each answer in the companion book. Grades above 80% should be considered passing.

Good Luck

TABLE OF CONTENTS

TEST A

Directions: Circle the correct answer choice

1.) If the total pump count is 235, the reporting time is 6 months, and the total number of repairs is 26, what is the mean time between repairs?
a) 45.2 months
b) 12.2 months
c) 54.2 months
d) 94.2 months

2.) Fill in the blank: "This equipment _____. take care of it!!"
a) Powers your process
b) Is fragile
c) Is unspared
d) Pays your salary

3.) An effective strategy for improving your plant's MTBF is to begin by addressing _____.
a) Low efficiency pumps
b) Bad actors
c) Self priming pumps
d) High profile

4.) A processing facility cannot exist if it is not_____.
a) Able to operate past the maximum design rates
b) Safe, environmentally friendly and profitable
c) A state of the art design
d) None of the above

5.) A centrifugal pump's operating point is determined by the intersection between the head-flow curve and the _____.
a) Efficiency curve
b) System curve
c) Power curve
d) None of the above

6.) A pump can generate 240 feet of head. If the specific gravity of the pumpage is 0.85 and you have 60 feet of suction pressure, what discharge pressure should you expect?
a) 110.4 psi
b) 88.3 psi
c) 129.8 psi
d) none of the above

7.) Bernoulli's principle states that the _____.
a) Flow is always conserved in a fluid stream
b) Pressure is always conserved in a fluid stream
c) Energy is never conserved in a fluid stream
d) Energy is always conserved in a fluid stream

8.) One key disadvantage of centrifugal pumps is that their _____ are typically less than positive displacement pumps.
a) Flows
b) Installation foot prints
c) Efficiencies
d) None of the above

9.) The MTBR curve peaks near the _____ flow.
a) Minimum
b) Maximum
c) Most laminar
d) Best Efficiency Point

10.) _____ is a damaging phenomenon occurring at the pump's impeller eye resulting from the collapse of vapor bubbles.
a) Cavitation
b) Erosion
c) Catalytic Reduction
d) Corrosion

11.) What does a centrifugal pump experience as it moves away from its best efficiency point?
a) Higher vibration
b) Greater shaft deflection
c) Higher pressure pulsations
d) All the above

12.) The three basic impeller types are:
a) Solid, porous, and enclosed
b) Enclosed, semi-open, and open
c) Solid, semi-open, and enclosed
d) Cast, enclosed, and open

Table 3.1, Summary of the 40/60/80 Rule

	<10 hp	<200 hp	>200 hp
Ample NPSHa or Suction head	40	60	80
Marginal NPSHa or suction head)	60	70	80

13.) If you have a pump with a 250 HP driver and a best efficiency point flow of 650 gpm, how low of a flow can you safely allow without harming the pump, based on Table 3.1?
a) 390 gpm
b) 325 gpm
c) 65 gpm
d) 520 gpm

14.) What are the two basic types of bearings?
   a) Spindle and journal bearings
   b) Rolling element and journal bearings
   c) Bushings and grease bearings
   d) none of the above

15.) The three (3) types of suction arrangements are:
   a) Netted suction, suction lift, and self-priming
   b) Flooded suction, suction lift, and self-priming
   c) Flooded suction, suction cooling, and self-priming
   d) Flooded suction, suction lift, and self-pressuring

Figure 4.8

16.) What pump type is shown in Figure 4.8?
   a) Multistage pump
   b) Multistage vertical pump
   c) Liquid ring pump
   d) Single stage overhung pump

17.) Which of these is not a type of pump seal?
   a) Trapping Seal
   b) Mechanical Seal
   c) Labyrinth Seal
   d) Packing

18.) _____ are considered the "bad boys" of all pump components.
   a) Bearings
   b) Mechanical seals
   c) Impellers
   d) Shafts

19.) If you have purchased a pump that is to pump an environmentally unfriendly liquid, what type of API plan should you consider?
   a) 52 (dual seals with unpressurized buffer fluid tank)
   b) 53A (dual seals with pressurized barrier fluid tank)
   c) 74 (dual gas seals with externally supplied barrier gas)
   d) Any of the above

API Seal Flush Plan 11

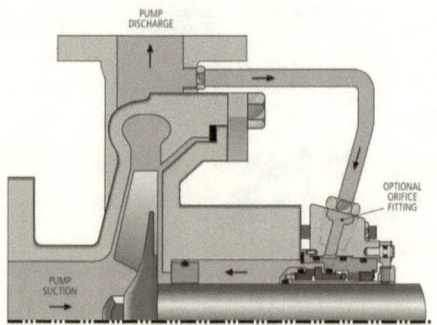

20.) True or False: The Plan 11 seal flushing arrangement does not required any action by the operator.
   a) True
   b) False

API Seal Flush Plan 32

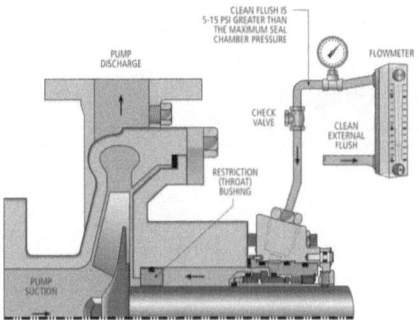

21.) The primary purpose of an API plan 32 is to_____.
    a) Improve pumping efficiency
    b) Keep the seal air free
    c) Expel solids from the sealing chamber
    d) Isolate the mechanical seal from the process

22.) _____ causes the insulation materials within the motor to degrade and eventually fail.
    a) Excessive heat
    b) Excessive vibration
    c) Normal vibration
    d) None of the above

23.) _____ laws allow you to estimate the effects of speed on flow, head, and power consumption.
    a) Speed
    b) Performance
    c) Approximation
    d) Affinity

24.) True or False: Operators should avoid directing high-pressure water wash near motor bearings.
    a) True
    b) False

25.) An "across the line" electric motor start means_____.
　　a) Full line voltage and current are connected to the motor during
　　　　start-up
　　b) Partial line voltage and current are connected to the motor
　　　　during start-up
　　c) Variable line voltage and current are connected to the motor
　　　　during start-up
　　d) None of the above

26.) _____ is a solid to semi-fluid mixture of a
　　thickening agent and liquid lubrication.
　　a) Oil
　　b) Grease
　　c) Additive
　　d) Chemical soap

27.) True or False: Grease is typically used in higher speed applications.
　　a) True
　　b) False

28.) True or False: Lubrication reduces friction between the rotating
　　and stationary components.
　　a) True
　　b) False

29.) True or False: It is normal for the oil level in a closed loop lubrication
　　system to rise during normal operation.
　　a) True
　　b) False

30.) The process that ensures that piping and pump casing are vapor or
　　air free is called _____.
　　a) Pressurizing
　　b) Cavitating
　　c) Normalizing
　　d) Venting

31.) Because product is used to lubricate internal bearings, extra care should be taken to _____ sealless pumps.
a) Vapor or air free
b) Align
c) Degauss
d) Balance

32.) If you suspect a pump is operating at too high a flow what can you try to remedy the situation?
a) Pinch the pump's discharge valve to create more backpressure
b) Open the pump's discharge valve to reduce backpressure
c) Increase the pumping temperature slightly
d) Raise the suction level

33.) The condition where a pump operates with insufficient or zero back-pressure is called _____.
a) Overpressure
b) Blowdown
c) Purging
d) Runout

34.) Your pump is discharging into an unpressurized header and has tripped on high amps several times. You are probably experiencing _____.
a) Runout
b) Overpressure
c) Purging
d) Blowdown

35.) The failure rate versus service time curve for a population of pumps is sometimes referred to as a _____ curve.
a) Bell
b) Bathtub
c) Pareto
d) "S"

36.) The most difficult start-up of a horizontal, flooded suction pump is _____.
a) After repair with an unpressurized system
b) A proven pump with an unpressurized system
c) A proven pump with a fully pressurized system
d) After repair with a fully pressurized system

37.) Which of these pump reliability statistics should be minimized?
a) Number of early failures per year
b) Pump inspection per year
c) Number of PM activities per year
d) All the above

38.) Before starting a flooded suction pump, always ensure_____.
a) The suction valve is partially closed
b) The suction level is at or above the recommended operating level
c) At least one low point bleeder is fully open
d) None of the above

39.) Submergence is the distance from a vertical pump's _____ to the suction level.
a) Suction bell
b) First bushing
c) Packing
d) Lower motor bearing

40.) If you see vortexes forming in the sump, what is probably the cause?
a) Strainer sump is plugged
b) Inadequate submergence
c) Pump is turning backwards
d) None of the above

41.) Which of these four (4) start-up scenarios are frequently encountered in vertical turbine pump start-up scenarios?
a) Unproven pump with an unpressurized system
b) Proven pump with an unpressurized system
c) Proven pump with a pressurized system

d) Unproven pump with a pressurized system

e) All the above

42.) Who should review your pump start-up procedures before implementation?
a) Plant manager
b) Process engineer
c) Machinery engineer
d) b&c
e) All the above

43.) The observable effects of a pump problem are called_____.
a) Causal factors
b) Side effects
c) Symptoms
d) Sympathetic effects

44.) The root of a pump problem is called _____.
a) A cause or causal factor
b) Side effect
c) Symptom
d) Sympathetic effect

45.) After a motor replacement, you are experiencing very low flows. What is the most likely cause?
a) Motor turning backwards
b) Too much back pressure
c) Wrong impeller
d) None of the above

46.) A plugged strainer can result in _____.
a) Cavitation
b) Low flow
c) Low power draw
d) All of the above
e) None of the above

47.) Reliability is a _____, not a destination.
   a) Process
   b) Philosophy
   c) Fad
   d) Journey

48.) A pump switching program applies to processes utilizing _____.
   a) Unspared pumps
   b) Multiple pump applications
   c) Main and spare pumps
   d) None of the above

49.) A process critical pump has failed numerous times in the past 6 months. What would be a prudent course of action?
   a) Form a root cause failure analysis team to determine the root cause
   b) Ignore the problem and see if it goes away
   c) Bring in the manufacturer to review your repair procedures
   d) Have your accounting office review all repair expenditures

50.) Which of these tools are useful for monitoring pumps in service?
   a) Pressure gauges
   b) Power monitor for electric motor
   c) Flow meter
   d) Vibration monitor
   e) All the above

TEST B

Directions: Circle the correct answer choice

1.) True or False: As a pump operator, your only job is to start and stop pumps.
   a) True
   b) False

2.) The worst 5 to 10 pumps based on the number of failures in a given time period are called _____.
   a) Repeaters
   b) Bad Applications
   c) Bad Selections
   d) Bad Actors

3.) An operator is burned trying to extinguish a fire caused by a product release that occurred because a seal leak was ignored and gradually worsened. Over $50,000 in damages was sustained by the pump, motor, and electrical wiring as a result of the fire. Which facet(s) of the operator's mission did the operator fail to accomplish in this situation?
   a) Protect life and limb
   b) Protect the environment
   c) Minimize the cost of ownership
   d) a and b
   e) All the above

4.) Which of these pump reliability statistics should be maximized?
   a) Mean time between early failures
   b) Mean time between catastrophic failures
   c) Mean time between repairs
   d) All the above

5.) For a pump with a given head capability, will the discharge pressure be greater or less when a heavier fluid is pumped?
   a) Greater
   b) Less

6.) Let's say you have a pump that is suppose to generate 462 ft of head at the rated flow. If you are pumping water at the rated flow and measure 160 psi of pressure across the pump, which of these statements is true?
a) You are producing more pressure than expected
b) You are producing the expected pressure
c) You are producing less pressure than expected
d) None of the above

7.) When any high velocity fluid stream slows down _____ is created.
a) Flow
b) Pressure
c) Cavitation
d) Heat

8.) NPSHr stands for _____.
a) Net positive sustainable pressure required
b) Net positive suction head required
c) Net preferred suction pressure required
d) None of the above

9.) True or False: If the manufacturer requires 20 feet of NPSHr to prevent cavitation and your process has 30 feet of NPSHa, cavitation will probably occur.
a) True
b) False

Figure 4.10

10.) What pump type is shown in Figure 4.10?
   a) Multistage pump
   b) Multistage vertical pump
   c) Liquid ring pump
   d) Single stage overhung pump

Figure 4. 11

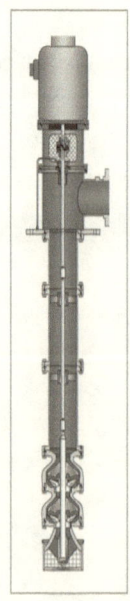

ROBERT X. PEREZ

11.) What pump type is shown in Figure 4.11?
  a) Multistage pump
  b) Multistage vertical pump
  c) Liquid ring pump
  d) Single stage overhung pump

12.) You have a pump that is taking suction from a vessel that is 10 feet above the pump's suction flange. From this you can conclude that the pump_____
  a) Has a suction lift
  b) Has a submerged suction
  c) Has a flooded suction
  d) Will always have ample of suction head

13.) An example of a sealless pump is a _____.
  a) Canned motor
  b) Centerline mounted
  c) Self-priming
  d) Fusion drive

14.) If you are planning to purchase a pump for a highly hazardous or toxic service, you should consider a _____ pump.
  a) Nuclear pump
  b) Sealless
  c) Triple sealed
  d) Centerline mounted

API Seal Flush Plan 21

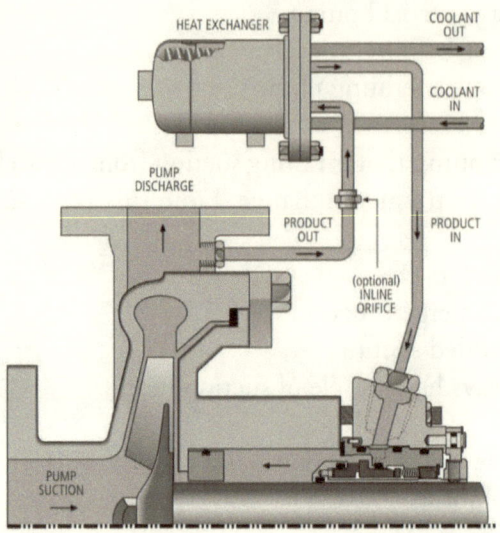

15.) Let's say you have a Plan 21 flush plan. What should you do first if you are not getting any cooling out of the cooler?
a) Check cooling water flow
b) Check product flow into cooler
c) Back flush cooler
d) Vent product lines

API Seal Flush Plan 53A

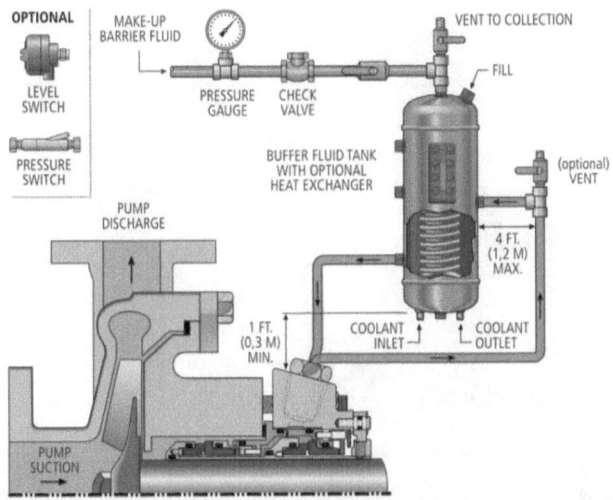

16.) If you have a pump with an API plan 53A for the seal, which of the following can you conclude?
a) You have dual seals
b) You have a seal pot
c) You have a pumping ring
d) All the above
e) None of the above

17.) Quenching is used when the sealed product has a tendency to
_____.
a) Set-up
b) Crystallize
c) Coke-up
d) All the above

18.) You should check the seal flush flow _____.
a) Every hour
b) Before start-up
c) After start-up
d) b and c

API Seal Flush Plan 53A

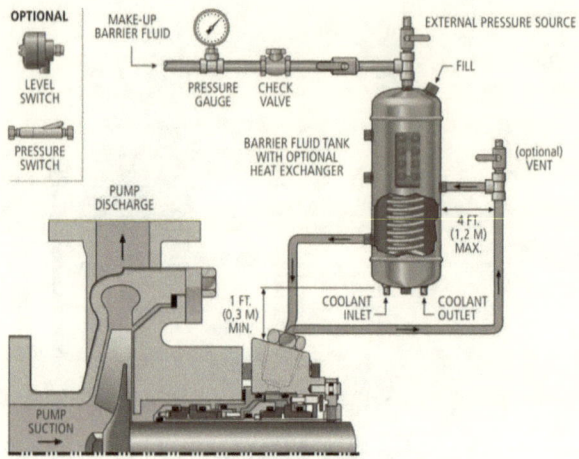

19.) If you have an API plan 53A with a gas blanket pressure of 80 psig and you know the maximum stuffing box pressure is 70 psig, what should you do?
   a) Lower the gas blanket pressure to 70 psig
   b) Raise the gas blanket flow to 70 scfm
   c) Nothing
   d) Raise the gas blanket pressure to 100 psig

20.) If you have a 400 HP motor, how long should you allow it to cool before restarting after 2 or 3 start attempts?
   a) One hour
   b) One day
   c) six hours
   d) No need to wait

21.) If you have 8" suction piping, how much straight piping should there be upstream of the pump's suction nozzle?
   a) 12" to 16"
   b) 32" to 80"
   c) 20" to 36"
   d) None

22.) A good quality motor should last a lifetime, and _____ of useful life is not uncommon even in harsh industrial plant environments.
a) 30 to 50 years
b) 3 to 5 years
c) 10 to 20 years
d) 30 to 50 months

23.) True or False: Over greasing can potentially damage electrical components of the motor.
a) True
b) False

24.) An oil's lubricating performance is greatly degraded by _____.
a) Reduction and egression
b) Contamination and ingression
c) Contamination and oxidation
d) Oxidation and reduction

25.) A lubrication system consisting of oil particles 1.0 to 3.0 microns in diameter, suspended in a current of air in a ratio of 1 part of oil to 200,000 parts of air is called _____.
a) Grease lubrication
b) Oil mist lubrication
c) Air mist lubrication
d) Splash lubrication

26.) You can keep you oil clean by _____.
a) Sealing, purging, and heating
b) Sealing, purging, and filtering
c) Sealing, venting, and filtering
d) None of the above

27.) Which of these is not a critical function of lubrication?
a) Reduce wear
b) Increase friction between the rotating and stationary components
c) Absorb shock
d) Minimize corrosion

28.) NPSHa stands for _____.
   a) Net positive sustainable pressure available
   b) Net positive suction head available
   c) Net possible suction pressure available
   d) None of the above

29.) To prevent cavitation you must ensure _____.
   a) NPSHa is less than the NPSHr
   b) NPSHa is equal to the NPSHr
   c) NPSHa is greater than the NPSHr
   d) None of the above

30.) The best means of eliminating air in suction piping and the pump casing prior to startups is with the use of:
   a) An air scavenging system
   b) High point vents
   c) Low point vents
   d) None of the above

31.) Multistage pumps should be vented _____.
   a) Daily
   b) By opening the suction and discharge valves all the way
   c) By opening all casing high points before start-up
   d) By opening all casing low points before start-up

32.) Which of these should you always check, if they apply, before start-ups? A) All oil lines B) All cooling water lines C) Seal flush line for proper flow D) Seal pot level E) Seal pot gas blanket
   a) A, B, and C
   b) A thru D
   c) C, D, and E
   d) All of the above

33.) The easiest start-up of a horizontal, flooded suction pump is _____.
   a) After repair with an unpressurized system
   b) A proven pump with an unpressurized system
   c) A proven pump with a fully pressurized system
   d) After repair with a fully pressurized system

34.) A pump fails less than 24 hours after it is repaired and started up. This represents a(n) _____ failure,
a) End of life
b) Pareto
c) Bathtub
d) Infant mortality

35.) A pump fails well past its design life. This represents a(n) _____ failure.
a) End of life
b) Infant Mortality
c) Bathtub
d) Pareto

36.) True or False: A single generic start-up procedure may be used for all pump start-ups.
a) True
b) False

37.) Vertical pumps with enclosed line shafts require _____.
a) External or lubricant flow to the line shaft bearings
b) Extra oil to the pump bearings
c) A slightly pinched discharge valve at all times
d) External or lubricant flow to the motor bearings

38.) Which of these factors are important to know when developing a start-up procedure?
a) Type of seal plan
b) If the pump's discharge is pressurized or unpressurized
c) If the pump operates above 200°F
d) All the above

39.) Which of these questions should you ask when developing a start-up procedure?
a) Do you have a flooded suction or suction lift?
b) Is there a discharge check valve?
c) a & b
d) None of the above

40.) What could cause you to experience too high of a flow?
   a) Wrong impeller diameter
   b) Wrong speed
   c) Not enough back pressure
   d) All the above

41.) The 3 most common flow symptoms are: Flow too low, flow too high, and _____.
   a) Bilateral flow
   b) Critical Flow
   c) Unstable flow
   d) None of the above

42.) After a planned unit outage, a vertical pump with a submerged suction bell is experiencing unsteady and low flows. What is the most probably cause?
   a) Inadequate submergence of the suction bell
   b) Too much submergence of the suction bell
   c) Operation at the best efficiency point
   d) None of the above

43.) After a motor replacement, you are experiencing lower than normal flows. What is the most likely cause?
   a) Wrong motor speed
   b) Too much back pressure
   c) Wrong impeller
   d) None of the above

44.) True or False: Unspared process pumps are generally more critical than spared one.
   a) True
   b) False

45.) Idle o-rings can fail when needed because they tend to _____.
   a) Slip
   b) Swell
   c) Decompose
   d) Take a set

46.) If you are pumping a product that tends to set-up or crystallize, it is wise to _____.
a) Always keep the spare cool to the touch to prevent polymerization
b) Flush the casing out with a process compatible fluid before placing it in stand-by service
c) Never shut down
d) None of the above

47.) A pump or set of pumps should be considered critical if a(n) _____.
a) Major failure could result in a major release, fire, or explosion
b) Undetected failure will result in a major (>50,000) repair cost
c) Major failure will result in a lengthy process outage
d) Any of the above

48.) After starting the spare pump and shutting down the main pump, you notice your flow drops to almost zero. What should you do?
a) Slightly pinch the spare pump's discharge valve as soon as possible
b) Pinch the main pump's discharge valve gradually to see if things stabilize
c) Completely close the main pump's discharge valve as soon as possible
d) Call your supervisor for guidance

49.) Which of these vibration levels is too high for continuous pump operation?
a) 0.02 inches per second
b) 0.15 inches per second
c) 0.5 inches per second
d) All of the above

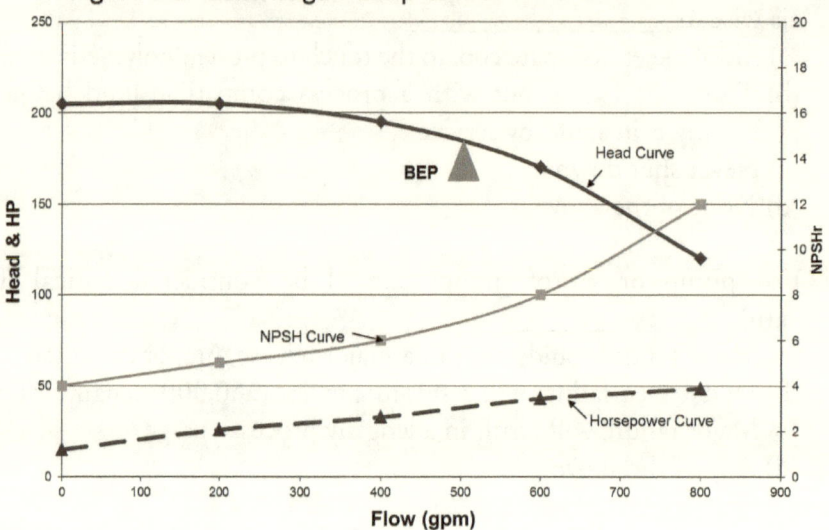

Figure 2.3- Centrifugal Pump Curve

50.) In figure 2.3 the BEP flow is at _____.
   a) 740 gpm
   b) 250 gpm
   c) 500 gpm
   d) None of the above

ROBERT X. PEREZ

TEST C

Directions: Circle the correct answer choice

1.) Extensive secondary damage and process release occurred because high vibration levels were ignored. What facet(s) of the operator's mission did the operator fail to accomplish in this situation?
a) Protect life and limb
b) Protect the environment
c) Minimize the cost of ownership
d) a and b
e) b and c

2.) A set of pumps is experiencing the following run intervals in months: 24, 22, 25, 30, 36, and 48. Is the pump's reliability improving or deteriorating?
a) Improving
b) Deteriorating

3.) If you find the MTBR for a set of pumps is deteriorating, what should you investigate?
a) Change in pump flow
b) Change in fluid properties
c) Change in preventative maintenance intervals
d) All the above

4.) How many failures would you experience in six months if you have a 1000 pumps and a MTBR of 50 months?
a) 120
b) 80
c) 50
d) 12

5.) How many failures would you experience in one year if you have a 200 pumps and a MTBR of 40 months?
a) 5
b) 40
c) 30
d) 60

6.) A pump can generate 231 feet of head. If you are pumping water (specific gravity of 1.0), what pressure rise should you expect across the pump?
a) 180 psi
b) 100 psi
c) 231 psi
d) none of the above

7.) True or False: If the manufacturer requires 30 feet of NPSHr to prevent cavitation and your process has 20 feet of NPSHa, cavitation will probably occur.
a) True
b) False

8.) If you find a pump cavitating, which of these will help eliminate the problem?
a) Increasing the liquid level
b) Cooling the product
c) Increasing the pressure in the suction vessel
d) All the above

9.) The pump manufacturer finds that its pump requires 20 feet of suction over the fluids vapor pressure at 500 gpm to prevent cavitation. From this you can say _____.
a) If you maintain more than 20 feet suction head you will not cavitate
b) If you maintain more than 20 feet suction head over the fluid's vapor pressure you will not cavitate
c) If you maintain more than 20 feet discharge head at all times you will not cavitate
d) None of the above

10.) NPSHr is an important pump parameter that tells you _____.
a) How much discharge pressure is needed to suppress cavitation
b) How much suction pressure is needed to suppress cavitation
c) How much flow is needed to suppress cavitation
d) None of the above

Table 3.1, Summary of the 40/60/80 Rule

	<10 hp	<200 hp	>200 hp
Ample NPSHa or Suction head	40	60	80
Marginal NPSHa or suction head)	60	70	80

11.) If you have a pump with a 150 HP driver, a best efficiency point flow of 300 gpm, and ample NPSHa, how low of a flow can you safely allow without harming the pump, based on Table 3.1?
a) 180 gpm
b) 240 gpm
c) 30 gpm
d) 120 gpm

12.) For a self-priming pump to work properly you must first
_____.
a) Align the pump to the motor
b) Air free the priming chamber
c) Check to see all bolts are properly torqued
d) Fill the priming chamber with water

13.) Which of these is not a typical pump component?
a) Agitator
b) Impeller
c) Shaft
d) Bearing

14.) Which of these is not a typical seal component?
a) O-ring
b) Spring
c) Gland
d) Modulator

Figure 4.4

15.) Is the bearing in Fig 4.4 a rolling element bearing or a journal bearing?
a) Rolling element bearing
b) Journal bearing

16.) In a flooded suction pump installation, the suction liquid level is _____ than the pump suction flange.
a) Higher
b) Lower

17.) In a self-priming pump installation, the suction liquid level is _____ than the pump suction flange.
a) Higher
b) Lower

18.) An example of a sealless pump is a _____.
a) Centerline mounted
b) Self-priming
c) Fusion drive
d) Magnetic drive

19.) True or False: When troubleshooting, is it vital to know the actual seal cavity pressure.
   a) True
   b) False

API Plan 12

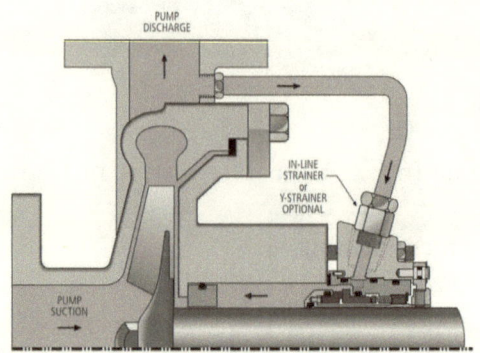

20.) If a seal with an API plan 12 fails, what should you do before starting up again?
   a) Check the seal strainer
   b) Change the o-ring material
   c) Check the pump to motor alignment
   d) None of the above

API Plan 52

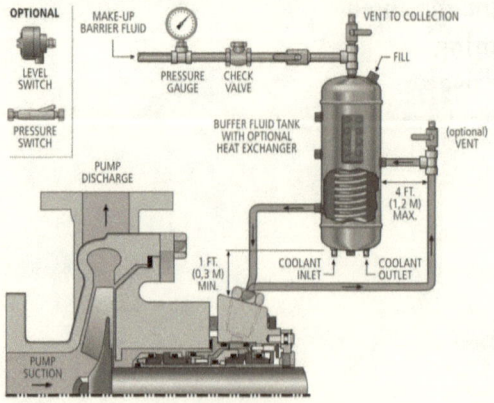

ROBERT X. PEREZ

21.) If you have an API plan 52 and your inner seal begins to leak, what should you see in the seal pot?
a) Seal pot pressure will drop
b) Seal pot level will rise
c) Seal pot level will decrease
d) None of the above

22.) Which of these can effect mechanical seal life?
a) Vibration
b) Pumping Temperature
c) a and b
d) Neither a nor b

23.) True or False: The fluid level in a seal pot used in an API plan 52 (shown above) can increase and decrease.
a) True
b) False

API Plan 53A

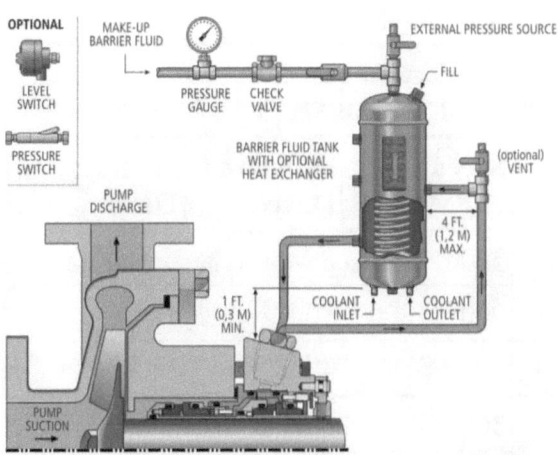

24.) The pressure on the gas blanket in an API plan 53A should be _____ over the maximum seal cavity pressure.
a) 5 psi   c) 2 psi
b) 15 psi   d) 30 psi

API Plan 74

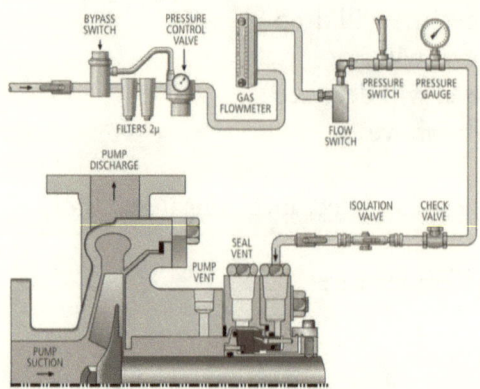

25.) The gas barrier pressure in an API plan 74 should be set to the _____.

a) Seal supplier's recommendations
b) Seal cavity pressure
c) Suction pressure plus 5 psi
d) None of the above

Frequency of Lubrication based on Speed, Frame, and Type of Service				
Frame	Speed (RPM)	Standard Duty	Severe Duty	Harsh Duty
143T-256T	3600	8 Mos.	4 Mos.	1 Mos.
	900 – 1800	30 Mos.	12 Mos.	4 Mos.
284T-365T	3600	8 Mos.	4 Mos.	1 Mos.
	900 – 1800	24 Mos.	12 Mos.	4 Mos.
404T- 447T	3600	8 Mos.	4 Mos.	1 Mos.
	900 – 1800	18 Mos.	8 Mos.	3 Mos.

ROBERT X. PEREZ

Notes:

1. When re-lubricating roller bearings divide the monthly service by two.
2. Standard duty = 8 hrs/day operation, light or normal loading, clean dust free environment.
3. Severe duty = 24 hrs/day operation, light to normal shock loading and vibration, exposure to dirty and dusty environment.
4. Harsh duty = 24 hrs/day operation, high ambient temperatures, normal to high shock loading, high vibrations, exposure to dirty and dusty environments.

26.) How often do you need to re-grease a Frame 404T (3600 rpm) in severe duty? (Refer to Table 6.2)
a) Every 4 months
b) Every 8 months
c) Every 16 months
d) Every 4 years

27.) Which of these is not an accepted greasing tip?
a) Keep grease gun out of the weather
b) Warm up grease gun to 200° F before use
c) Don't over grease
d) Ensure grease coupler is clean before use

28.) If you experience a rapid bearing failure after changing grease suppliers, what should you do?
a) Check grease compatibility of the new grease with the old grease
b) Nothing. Hope it was a fluke
c) Pump in more grease next time
d) Try a heavier grease next time

29.) True/False: If you are taking pumps out of service, it is OK to pinch off the oil mist lines that will not be needed.
a) True   b) False

30.) You should check _____ on all wet sump applications daily.
a) oil level
b) oil condition
c) viscosity
d) a and b

31.) Before starting, you should always preheat pumps that operate above _____.
a) 500°F
b) 32°F
c) -150°F
d) 150°F

32.) Which of these statements is true for parallel pump operations.
a) Multiple pumps operating in parallel will always pump the same flow
b) Multiple pumps operating in parallel will never pump the same flow
c) Multiple pumps operating in parallel should be monitored to ensure they are pumping at acceptable flows
d) Pumps of different types and speed are allowed to operate in parallel

33.) If you don't have enough back pressure on a pump, which statement will be true? (Assume you have a typical radial flow pump.)
a) Pressure will be higher than normal
b) Flow will be lower than normal
c) Flow will be higher than normal
d) Power will be lower than normal

34.) True or False: Pumps with the same rated flows can always be operated in parallel.
a) True
b) False

35.) Which of these conditions is best for your pump?
   a) NPSHa > NPSHr
   b) NPSHa = NPSHr
   c) NPSHa < NPSHr
   d) NPSHa = 0

36.) If you have just repaired a pump, what should you do before putting it back in service?
   a) Vent the casing
   b) Vent the suction piping
   c) Fully open the suction valve
   d) All the above

37.) After a major unit outage where all vessels and piping has been de-inventoried, but no pump work has been done, what start-up procedure would you use on each pump?
   a) After repair with an unpressurized system
   b) A proven pump with an unpressurized system
   c) A proven pump with a fully pressurized system)
   d) After repair with a fully pressurized system

38.) After a major repair and electric motor replacement, you are ready for start-up, but you are not sure if the discharge header is fully pressurized. What start-up procedure should you use?
   a) After repair with an unpressurized system
   b) A proven pump with an unpressurized system
   c) A proven pump with a fully pressurized system
   d) After repair with a fully pressurized system

39.) After startup you should monitor the pump _____ to ensure it is pumping and operating correctly.
   a) Every hour for the next 24 hours
   b) Every 15 minutes for an hour
   c) Every minute for the first hour
   d) No need to stay after the pump is started

40.) After your pump reaches full operating speed, you should_____.
a) Look at all pressure gauges to ensure normal and steady pressures
b) Feel the pump to ensure pump vibrations are normal
c) Listen for unusual noises, such as rubbing
d) All the above

41.) You are preparing to start a vertical pump with a submerged suction. You notice you actually have two (2) feet less submergence that recommended. What should you do?
a) Nothing. Go ahead and start-up
b) Wait until you have the required submergence
c) Start-up but make sure you discharge valve is pinched slightly
d) None of the above

42.) A foreign object lodged in the impellers will probably lead to _____.
a) Lower vibration
b) Higher flow
c) Higher vibration
d) Higher discharge pressure

43.) What symptoms would you expect to see if an impeller with a smaller diameter than the design diameter was installed?
a) Lower pressure and higher flow
b) Lower pressure and lower flow
c) Higher pressure and lower flow
d) Higher pressure and higher flow

44.) If your pump flow drops, which of the following causes is the most likely?
a) Plugged suction strainer
b) Lower back pressure
c) Larger impeller diameter was accidentally installed
d) None of the above

45.) Which of these individuals would make the best reliability sponsor?
a) Project engineer
b) Vibration technician
c) Process engineer
d) Maintenance manager

46.) When you are switching from the main pump to its spare make sure you _____.
a) Start the spare and allow both to run to ensure the spare is healthy before shutting down the main
b) Shut the main pump down immediately after starting the spare to avoid over pressuring the system
c) Shut both pumps down momentarily to check both check valves before starting the spare
d) None of the above

47.) Which of these is a vital requirement when participating in a root cause failure analysis?
a) Try to draw a conclusion as soon as possible
b) Always keep an open mind
c) Try to find blame
d) None of the above

Figure 1.2

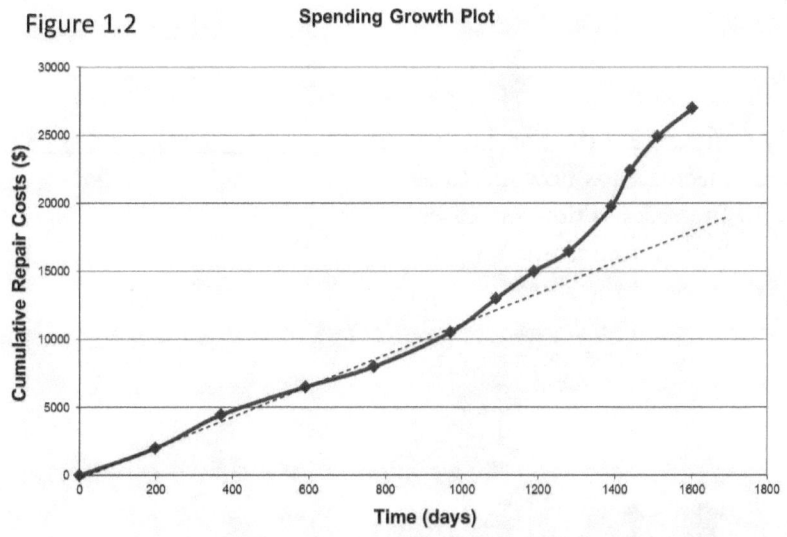

Figure 1.2   Spending Growth Plot

48.) What is Figure 1.2 telling you about the cost of maintenance for this pump?
   a) Maintenance costs per period of time are increasing
   b) Maintenance costs per period of time are decreasing
   c) Maintenance costs per period of time are constant

Figure 2.3

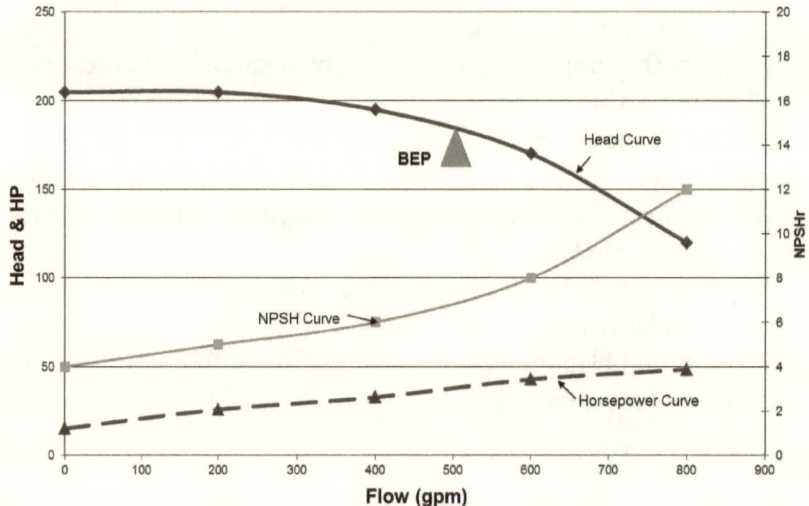

Figure 2.3- Centrifugal Pump Curve

49.) In figure 2.3 the horsepower curve _____.
   a) Decreases as flow increases
   b) Increases as flow increases
   c) Remains constant
   d) None of the above

ROBERT X. PEREZ

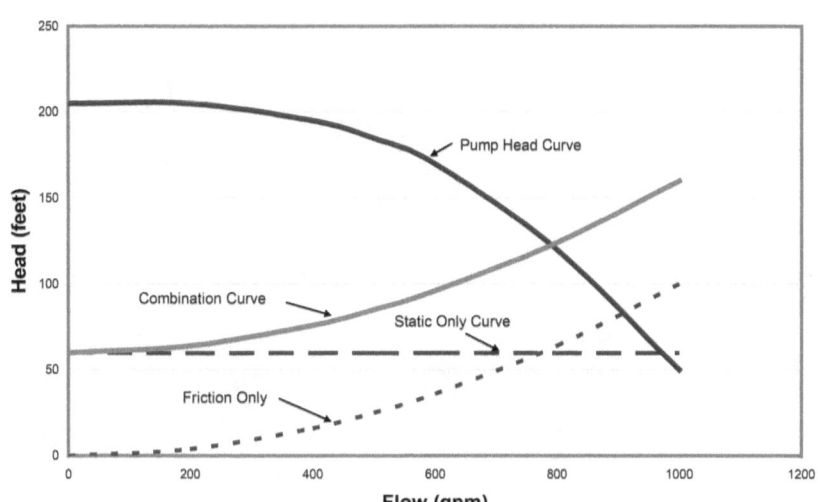

Figure 2.4. Static, Friction, and Combination System Curves

50.) In figure 2.4, which curve represents a system with a constant static head and fluid friction head?
   a) Combination curve
   b) Static only curve
   c) Friction only curve
   d) None of the above

ANSWERS TO
TESTS A, B, & C

OPERATOR'S GUIDE TO CENTRIFUGAL PUMPS—TEST A ANSWERS

1.) c) 54.2 months

$$MTBR = \frac{M \times T}{R} = \frac{235 \times 6}{26} = 54.23 \text{ months between repairs}$$

Chapter 1, page 15

2.) d) Pays your salary
Chapter 1, page 13

3.) b) Bad actors
Chapter 1

4.) b) Safe, environmentally friendly and profitable
Chapter 1

5.) b) System curve
Chapter 2, page 26

6.) a) 110.4 psi

$$\frac{(240\,ft + 60\,ft) \times 0.85}{2.31} = 110.4\,psi$$

Chapter 2, page 22

7.) d) Energy is always conserved in a fluid stream
Chapter 2, page 28

8.) c) Efficiencies
Chapter 2

9.) d) Best Efficiency Point
Chapter 3, page 32

10.) a) Cavitation
Chapter 3, page 35

11.) d) All the above
Chapter 3

12.) b) Enclosed, semi-open, and open
Chapter 4, page 41

13.) d) 520 gpm

$$650\,gpm \times 0.8 = 520\,gpm$$

Chapter 3, page 33

14.) b) Rolling element and journal bearings
Chapter 4, page 41

15.) b) Flooded suction, suction lift, and self-priming
Chapter 4

16.) d) Single stage overhung pump
Chapter 4, page 45

17.) a) Trapping Seal
Chapter 4

18.) b) Mechanical Seals
Chapter 5, page 52

19.) d) Any of the above
Chapter 5

20.) a) TRUE
Chapter 5, page 65

21.) d) Isolate the mechanical seal from the process
Chapter 5

22.) a) Excessive heat
Chapter 6, page 80

23.) d) Affinity
Chapter 6, page 82

24.) a) True
Chapter 6, page 78

25.) a) Full line voltage and current are connected to the motor
Chapter 6

26.) b) Grease
Chapter 7, page 84

27.) b) False. Grease is typically used in lower (1800 rpms or less) speed applications.
Chapter 7, page 85

28.) a) True
Chapter 7, page 84

29.) b) False
Chapter 7

30.) d) Venting
Chapter 8, page 91

31.) a) air or vapor free
Chapter 8, page 91

32.) a) Pinch the pump's discharge valve to create more backpressure
Chapter 8

33.) d) Runout
Chapter 8, page 96

34.) a) Runout
Chapter 8, page 96

35.) b) Bathtub
Chapter 9, page 104

36.) a) After repair with an unpressurized system
Chapter 9

37.) a) Number of early failures per year
Chapter 9, page 104

38.) b) The suction level is at or above the recommended operating
level
Chapter 9

39.) a) Suction bell
Chapter 10, page 116

40.) b) Inadequate submergence
Chapter 10

41.) e) All the above
Chapter 10

42.) d) b&c
Chapter 10, page 121

43.) c) symptoms
Chapter 11, page 123

44.) a) cause or causal factor
   Chapter 11, page 123

45.) a) Motor turning backwards
   Chapter 11

46.) d) All of the above
   Chapter 11, page 127

47.) d) Journey
   Chapter 12, page 129

48.) c) Main and spare pumps
   Chapter 12, page 134

49.) a) Form a root cause failure analysis team
   Chapter 12

50.) e) All the above
   Chapter 12, page 131

OPERATOR'S GUIDE TO CENTRIFUGAL PUMPS—TEST B ANSWERS

1.) b) False
Chapter 1

2.) d) Bad Actors
Chapter 1, page 18

3.) e) All the above
Chapter 1

4.) d) All the above
Chapter 1, page 18

5.) a) Greater
Chapter 2

6.) c) You are producing less pressure than expected
The pump should be generating $\frac{462\,feet}{2.31} = 200\,psi$
Chapter 2, page 22

7.) b) Pressure
Chapter 2, page 20

8.) a) Net positive suction head required
Chapter 3, page 35

9.) b) False
Chapter 3, page 37

10.) a) Multistage pump
Chapter 4, page 46

11.) b) Multistage vertical pump
Chapter 4, page 46

12.) c) Has a flooded suction
Chapter 4

13.) a) Canned motor
Chapter 4, page 49

14.) b) Sealless
   Chapter 4, page 49

15.) a) Check cooling water flow
   Chapter 5

16.) d) All the above
   Chapter 5

17.) d) All the above
   Chapter 5

18.) d) b and c
   Chapter 5

19.) d) Raise the gas blanket pressure to 100 psi
   Chapter 5

20.) a) One hour
   Chapter 6, page 80

21.) b) 32" to 80"
   $8" \times 4 = 32" - to - 8" \times 10 = 80"$

   Chapter 6, page 100

22.) a) 30 to 50 years
   Chapter 6

23.) a) True
   Chapter 6, page 78

24.) c) contamination and oxidation
   Chapter 7, page 86

25.) b) Oil mist lubrication
   Chapter 7, page 86

26.) b) Sealing, purging, and filtering
   Chapter 7, page 89

27.) b) Increase friction between the rotating and stationary components
   Chapter 7, page 84

28.) b) Net positive suction head available
   Chapter 8, page 94

ROBERT X. PEREZ

29.) c) NPSHa is greater than the NPSHr
    Chapter 8, page 94

30.) b) High point vents
    Chapter 8, page 91

31.) c) By opening all casing high points before start-up
    Chapter 8

32.) d) All A thru E
    Chapter 9, page 106

33.) c) A proven pump with a fully pressurized system
    Chapter 9, page 111

34.) d) Infant mortality
    Chapter 9, page 104

35.) a) End of life
    Chapter 9, page 104

36.) b) False
    Chapter 10, page 120

37.) a) External or lubricant flow to the line shaft bearings
    Chapter 10, page 115

38.) d) All the above
    Chapter 10, page 120

39.) c) a & b
    Chapter 10, page 120

40.) d) All the above
    Chapter 11

41.) c) Unstable flow
    Chapter 11

42.) a) Inadequate submergence of the suction bell
    Chapter 11, page 127

43.) a) Wrong motor speed
    Chapter 11, page 127

44.) a) True
    Chapter 12

45.) d) Take a set
Chapter 12

46.) b) Flush the casing out with a process compatible fluid before
placing it in stand-by service
Chapter 12

47.) d) Any of the above
Chapter 12

48.) c) Completely close the main pump's discharge valve as soon
as possible
Chapter 12

49.) c) 0.5 inches per second
Chapter 12, page 133

50.) c) 500 gpm
Chapter 2, page 24

OPERATOR'S GUIDE TO CENTRIFUGAL PUMPS—TEST C ANSWERS

1.) e) b and c
Chapter 1

2.) a) Improving
Chapter 1

3.) d) All the above
Chapter 1

4.) a) 120

$$MTBR = \frac{M \times T}{R} \Rightarrow R = \frac{M \times T}{MTBR} = \frac{1000 \times 6}{50} = 120\,repairs$$

Chapter 1, page 15

5.) d) 60

$$MTBR = \frac{M \times T}{R} \Rightarrow R = \frac{M \times T}{MTBR} = \frac{200 \times 12}{40} = 60\,repairs$$

Chapter 1, page 15

6.) b) 100 psi

$$\frac{\ddot{u}\quad feet}{2.31\,feet\,/\,psi} = \ddot{u} \quad \ddot{u}$$

Chapter 2, page 22

7.) a) True
Chapter 3, page 37

8.) d) All the above
Chapter 3

9.) b) If you maintain more than 20 feet suction head over the fluid's vapor pressure you will not cavitate
Chapter 3, page 35

10.) b) How much suction pressure is needed to suppress cavitation
Chapter 3

11.) a) 180 gpm

$$300\,gpm \times 0.60 = 180\,gpm$$

Chapter 3, page 33

12.) d) Fill the priming chamber with water
Chapter 4, page 49

13.) a) Agitator
Chapter 4, page 40

14.) d) Modulator
Chapter 4, page 43

15.) a) Rolling element bearing
Chapter 4, page 41

16.) a) Higher
Chapter 4, page 47

17.) b) Lower
Chapter 4, page 49

18.) d) Magnetic drive
Chapter 4, page 49

19.) a) True
Chapter 5, page 58

20.) a) Check the seal strainer
Chapter 5, page 66

21.) b) Seal pot level will rise
Chapter 5

22.) c) a and b
Chapter 5

23.) a) True
Chapter 5

24.) d) 30 psi
Chapter 5, page 74

25.) a) Seal supplier's recommendations
Chapter 5, page 76

26.) a) Every 4 months
Chapter 6, page 79

27.) b) Warm up grease gun to 200°F before use
Chapter 7, page 85

ROBERT X. PEREZ

28.) a) Check grease compatibility of the new grease with the old grease
Chapter 7

29.) b) False
Chapter 7, page 87

30.) d) a and b
Chapter 7, page 89

31.) d) 150°F
Chapter 8, page 94

32.) c) Multiple pumps operating in parallel should be monitored to ensure they are pumping at acceptable flows
Chapter 8, page 99

33.) c) Flow will be higher than normal
Chapter 8

34.) b) False
Chapter 8

35.) a) NPSHa > NPSHr
Chapter 8

36.) d) All the above
Chapter 8

37.) b) A proven pump with an unpressurized system
Chapter 9

38.) a) After repair with an unpressurized system
Chapter 9

39.) b) Every 15 minutes for an hour
Chapter 9, page 109

40.) d) All the above
Chapter 10, page 115

41.) b) Wait until you have the required submergence
Chapter 10

42.) 42.) c) Higher vibration
Chapter 11

43.) b) Lower pressure and lower flow
   Chapter 11

44.) a) Plugged suction strainer
   Chapter 11, page 127

45.) d) Maintenance manager
   Chapter 12

46.) a) Start the spare and allow both to run to ensure the spare is healthy before shutting down the main.
   Chapter 12

47.) b) Always keep an open mind
   Chapter 12, page 137

48.) a) Maintenance costs per period of time are increasing
   Chapter 1, page 17

49.) b) Increases as flow increases
   Chapter 2, page 248.)

50.) a) Combination curve
   Chapter 2

INDEX

R

S

T

V

W

www.ingramcontent.com/pod-product-compliance
Lightning Source LLC
Chambersburg PA
CBHW030848180526
45163CB00004B/1490